Captain John Kidwell was the first person to introduce the "Smooth Cayenne" variety of pineapple commercially in Hawaii.

History Of Pineapple In Hawaii

No one knows exactly when pineapple was first brought to these islands. Since pineapple was often carried along on voyages to prevent scurvy, som[e] believe that a wayward Spanish galleon traveling the Manila-Acapulco route was responsibl[e] for accidentally introducing pineapple to Hawaii long befo[re] Captain James Cook's arrival i[n] 1778.

The first recorded planting was on January 21, 1813, by Francisco de Paula y Marin, a Spanish horticulturist and advisor to King Kamehameha. However, modern Hawaiian pineapple had its beginning in 1882 when Captain John Kidwell imported 1,000 slips o[f] the "Smooth Cayenne" cultiva[r] from South America. About 600 survived and reached maturity on the small plantations he started on Oahu. In 1892 he began canning operations and the first major attemp[t] at serious production was underway.

Kidwell's operations faltere[d] however, and it was a young Harvard graduate, James D. Dole, who eventually established the first successful pineapple plantation in Hawaii.

James Drummond Dole, whose name is still associated with pine apple today, was the first person t[o] foresee the huge market possibilit[y] for pineapples in the world.

To the right: Oahu pineapple fiel[d]

In 1901, with $20,000 and 12 acres in Wahiawa, Dole established the Hawaiian Pineapple Company, later named the Dole Company, now a division of Castle & Cooke, Inc. The initial harvest in 1903 yielded 1,893 cases of canned fruit. By 1905, production reached 25,000 cases and 20 years later, 2.8 million cases.

Dole, however, was not the only successful pioneer. In 1898 a group of California farmers arrived in Hawaii. Among them was Alfred W. Eames, who founded the Hawaiian Islands Packing Company in 1906, and built a cannery in Wahiawa. Eleven years later the operation became part of the California Packing Corporation, now known as the Del Monte Corporation.

On Maui, the Baldwin family began pineapple operations as early as 1903. Nine years later they formed Baldwin Packers which in 1962 was merged into the Maui Pineapple Company, which has grown pineapples since 1906.

By 1940, Hawaii had become the world's largest grower of pineapple, accounting for 80 percent of the world's supply. Cultivation peaked around the late 1950's and early 1960's, and the 1986 crop was valued at $90 million.

A researcher inspects a field of pineapple.

Aerial view of Maui pineapple fields.

Lanai, the Pineapple Island

The story of pineapple in Hawaii is not complete without a chapter on Lanai, the "Pineapple Isle." According to legend, Lanai was the home of evil spirits and it was shunned by the natives until a young prince drove them away, making the island safe for mortals. But the early people still regarded it as too inhospitable and used it as a place of exile for women offenders.

Mormon elders and ranchers attempted to settle the island in the mid 1800's, but failed, primarily because of lack of water. With no water and much of the land denuded of trees by wild sheep, goats and cattle, Lanai was viewed as a wasteland, and in 1920 the population dropped to less than 100.

In 1902, Charles Gay and his family arrived on Lanai. His dream of acquiring all of the government lands and converting them into fee simple was a long and frustrating process. However, with the help of Cecil Brown, president of the First National Bank, the Lanai Company was established, and Gay purchased all but 600 acres of the island.

Left: Lanai pineapple fields seen from Hii Bench area. The spot is named "Director's Point" due to the fact that James Dole and his staff used to inspect the development of the plantation from this spot.

With hopes of building a sugar industry on the island, the Lanai Company planted trial plots of sugar beets. But, lack of water spelled doom for yet another venture. The company was then purchased by Frank and Harry Baldwin who concentrated on ranching. Meanwhile, Charles Gay formed Gay Ranch and started growing pineapple under contract with Haiku Pineapple Company on Maui.

In 1922, the Hawaiian Pineapple Company under the enlightened leadership of James Dole purchased the Baldwin properties on Lanai, and later the Gay Ranch, thereby acquiring essentially the entire island. After an additional investment of $1 million for water development, the Hawaiian Pineapple Company began its Lanai operations in 1924 on 300 acres. Hard work lay ahead with land choked with cacti and the constant tradewinds sweeping over the island. Paved roads were built and a safe harbor at Kaumalapau was blasted out of a sheer cliff to haul the fruit to Honolulu. In 1925, another 900 acres were planted. The following year a 3.3 million-gallon reservoir was built with a $1 million investment in water development. The first harvest came in 1926 and the population swelled to 1,000 in the next year.

The population tripled in the next two decades and Lanai became the largest pineapple plantation in the world, accounting for nearly half of all the pineapple processed by Dole. To accommodate the large labor force and their dependents, good schools, playgrounds, and recreation and health facilities were built. Pineapple so pervaded the way of life that school days were arranged so that children were free to help with the harvest.

Before 1940, pineapples are all harvested by hand as seen in this early plantation shot.

The Growing Of Pineapple

Known scientifically as Ananas comosus (L.) Merr., the pineapple is a perennial herb belonging to the Bromeliaceae, a family almost exclusively restricted to the New World tropics. Most of the species are epiphytes, often living high up in the canopies of giant forest trees, absorbing water and nutrients not through the roots which merely serve to anchor the plants to their supports, but through microscopic hairs at the base of the leaves.

The bluish flowers of the pineapple plant are borne in compact terminal heads sur–rounded at the base by reddish bracts.

The natural flowering time of a pineapple is in mid-December and fruiting follows in the summer. Today pineapple planters can induce flowering artificially during different times of the year to make harvesting possible all year-round.

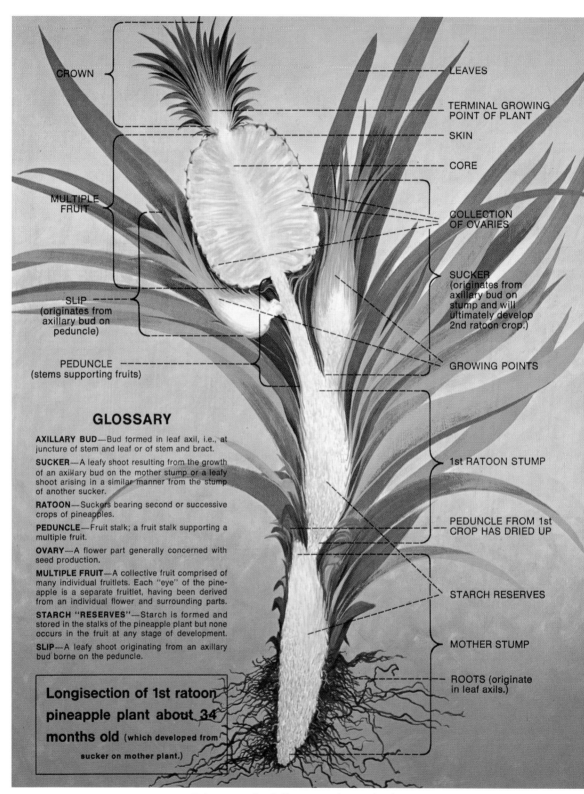

Diagram of a full-grown pineapple plant.
Average size of a plant today is usually 3 feet tall.

he fruit, formed without fertilization of the flowers, is technically a multiple or collective fruit formed by the coalescence of 100 to 200 individual berrylike fruitlets or "eyes." At the center is a porous core which is part of the peduncle, or stalk, of the flower. At the top of the fruit is a leafy crown.

After fruiting, growth continues in the side buds at the axils of the leaves. These buds grow into side branches which themselves bear fruit. This is called the ratoon crop. In this way, the plant will continue growing for 50 years. In commercial production, however, only one or two ratoon crops is harvested before the plants are uprooted. The branches emerging from the main stalk are called suckers and those emerging from the peduncle are called slips.

This young pineapple fruit will continue to grow 6 to 8 more weeks before it's ready to be harvested. As it gets closer to harvesting, the crown will take on a reddish tint, and the fruit's eyes will flatten.

Planting & Harvesting

The normal crop cycle between successive plantings is about four years. Between crop cycles the old plants are knocked down, crushed and plowed, leaving some of the old plant material in the soil. The fields are then graded for the next planting. Black polyethylene film strips for mulching are laid and the soil is fumigated to destroy nematodes and other harmful pests. After 48 hours, slips, shoots or crowns are planted by hand through the mulching film. The mulching film helps control weeds, increase soil temperature, conserve water and reduce leaching of soil nutrients. Between 22,000 and 26,000 plants are planted per acre. Irrigation is performed by a drip system laid beneath the film.

Pineapple workers' usual attire: Hat, bandanna, and long sleeves to protect against the sun, dust, and sharp spines on pineapple leaves.

To the right: Occasional manual weeding is necessary.

Field irrigator, developed on Lanai in the 50's, now used only right after planting.

Boxed fresh pineapples ready to be shipped all over the world.

A "Green Shell" pineapple ready to be harvested.

Overhead booms are used for the occasional spraying of iron sulfate to overcome nutritional deficiencies.

Flowers bloom in approximately 15 months and the first harvest occurs 18 to 22 months after planting. A year later the first ratoon crop is ready for harvest. Fields are manually

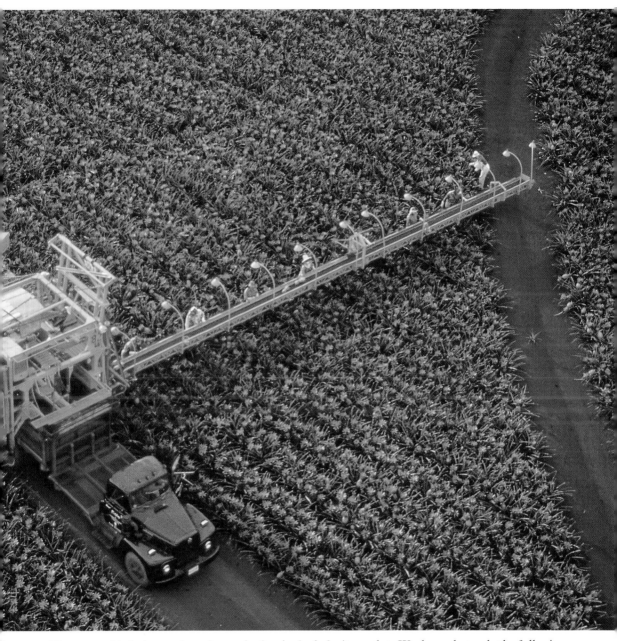

Conveyor Harvester aids in harvesting pineapples for the fresh-fruit market. Workers select only the fully ripe fruits to put on the conveyor belt of the machine to transport them to the harvester bin. The fruits are then packed carefully in the bins upside down by hand to avoid damage.

harvested several times during the season. Once harvested, fruit for canning will not become any sweeter, therefore only fully ripe fruit is picked. The fruit may be "green shell" ripe with little or no yellowing, or quite yellow when ripe. Pineapple ripens quickly and to prevent spoilage the fruit is rushed to the cannery and processed as rapidly as possible.

Fresh Hawaiian pineapple is available in most mainland and foreign markets within a few days of harvest. To avoid bruising, pineapples for the fresh-fruit market are stacked upside down on their crowns in harvester bins and transported to the packing center where they are washed, inspected, sorted by size and packed. Within hours they arrive at the docks or airport for shipment overseas.

Pineapples that are going to the cannery all had their crowns taken off in the field during harvesting.

Pineapples going through the automatic washing machine.

Canning

Soon after harvest the pineapples arrive at the cannery in specially designed bins.

Up to two million fruits can arrive at the cannery each day during peak season. The fruits are carried on conveyor belts into the cannery where they are washed and graded by size. Undersized fruits are usually processed into juice or crushed pineapple and other sizes are sent to specific packing lines.

The gateway to each line is an ingenious machine used in pineapple canneries all over the world and named after its inventor, Henry Ginaca of the Dole Pineapple Company. The Ginaca machine has circular revolving cylindrical shaped blades which cut the shell from the fruit; other mechanisms chop the ends off and core the fruit.

Ginaca machines at work.
The small picture shows the shelled pineapples
going through the trimming table before they enter the slicer.

Lines of gloved workers at the trimming tables during peak season. Sanitary clothing is required for all workers.

Lines of gloved workers do the final trimming before the fruit goes through a slicing machine. The slices are then sorted by more gloved workers.

In twenty minutes, the pineapple is completely processed, and its field-ripened quality is preserved.

Most popular of the canned pineapple products are the familiar round SLICES. Other specialty lines such as TIDBITS, SPEARS, CHUNKS, and CRUSHED PINEAPPLE are made from odd-size fruits unsuitable for slices. Bruised or crushed fruits, cores and broken slices are processed into juice.

Diced chunks of pineapple are being put into cans automatically.

Selecting Fresh Pineapple

To begin with, select a pineapple that is plump and fresh looking. Fresh, green leaves in the crown are a good sign. Ease in pulling the leaves out of the crown is not a sign of ripeness or good quality. The body should be firm—not soft. Shell color is not necessarily a sign of maturity or ripeness. A very green pineapple may be a ripe one; on the plantation this is called "green-shell ripe." Shell colors of ripe fruit are divided into seven groups or levels ranging from No.0, all green, to No.6, all yellow. Shell colors shown to the right are No.6.

Fresh pineapple from Hawaii is picked at maximum ripeness for delivery to U.S. and Canadian markets. The sooner it is eaten, the better.

Fresh pineapple contains bromelain, a protelytic enzyme that breaks down protein in a manner similar to what happens in digestion. For this reason:

- *Gelatin made with fresh pineapple will not set.*
- *Cottage cheese, sour cream and other dairy products should not be mixed with fresh pineapple until ready to be served.*

However, you can use fresh pineapple in a meat marinade to add a flavor accent and help tenderize the meat.

Pineapple, called "halakahiki" in Hawaiian (meaning foreign fruit) is also known as "the King of Fruits" because of its crown and taste. Shown here is the "Smooth Cayenne" variety.

Cutting Fresh Pineapple

Cutting the fruit in half is the first step in preparing fresh pineapple. It's easy; just use a long, sharp knife, start from the bottom and cut through the leafy crown.

Hollowing out the shell of the halved pineapple is the next step in preparing many colorful and attractive dishes. With a short, serrated knife, cut around the fruit one-half inch from the shell to make a serving bowl for a pineapple tuna salad or a hot chicken curry. Fresh pineapple can be cut in a variety of ways to add visual appeal to any occasion.

FOUR POPULAR CUTTING METHODS:

SPEARS:

- With a sharp knife, cut off the top and bottom of the pineapple. Then cut the shell away a strip at a time.
- Remove the "eyes" by cutting away diagonal strips.
- Cut the fruit cylinder into spears. To make smaller wedges, cuts spears crosswise.

QUARTERS:

- Cut the pineapple in half and then in quarters, cutting from the bottom through the crown with a sharp knife.
- Cut out the hard fibrous core. Leave the crown on for decoration.
- Loosen the fruit by cutting close to the rind with a curved knife.
- Cut crosswise through the fruit several times, then lengthwise once or twice to make bite-size pieces.

PINEAPPLE RUBY:

- With a sharp knife, cut off the top and bottom of the pineapple. Save the top and crown.
- Insert the knife close to shell and cut completely around the pineapple with a sawing motion.
- Remove the cylinder of the pineapple and cut into spears. Put the spears back together to form the cylinder again.
- Set the shell on the serving dish, and put the cylinder of spears back inside. Put the top and crown back on the pineapple to give it an uncut look.

OUTRIGGER:

- Quarter the pineapple, leaving the crown on.
- Loosen the fruit with a curved knife by cutting under and around the core, without removing the core. Then insert the knife and loosen the fruit from the shell.
- Remove the fruit and cut crosswise several times.
- Slip section of fruit back into the shell under the core and arrange in a staggered pattern.

HOW TO CUT A PINEAPPLE

Y CUT

Twist crown from pineapple.

Cut in half, then in quarters.

Trim off ends and core. Remove fruit from shell with a curved knife.

Cut into bite-size chunks.

PINEAPPLE BOAT

1. Cut pineapple in half lengthwise through crown.

2. Remove fruit from shell with curved knife, leaving shell intact.

3. Cut into quarters, remove core.

4. Cut into bite-size chunks.

OUTRIGGER

1. Quarter the pineapple, leaving the crown on.

2. Loosen fruit with a curved knife by cutting under and around the core, but don't remove the core.

3. Remove the fruit and cut crosswise several times.

4. Slip sections of fruit back into the shell under the core and arrange in a staggered pattern.

RECIPES – Drinks and Pupus

Teriyaki Chicken and Pineapple on Skewers, with a Mai Tai in the background.

TERIYAKI CHICKEN AND PINEAPPLE ON SKEWERS

1 pound boneless chicken breasts
stalks green onion, cut into 1 inch lengths
can (20 oz.) chunk pineapple in juice
1/2 cup soy sauce
tablespoon mirin (Japanese rice cooking wine)
tablespoons sugar
teaspoon minced ginger
clove garlic, minced

Remove skin from chicken breasts. Cut chicken into strips 1 inch wide; set aside. Drain pineapple juice, reserving 1/4 cup. Thread chicken, green onions and pineapple on bamboo skewers.

Combine the reserved pineapple juice, soy sauce, mirin, sugar, ginger and garlic; pour over chicken and marinate 30 minutes. Broil, basting frequently with marinade, until chicken is no longer pink inside, about 6 to 8 minutes. Makes 18 skewers.

MAI TAI

1 1/2 ounces light rum
1/2 ounce dark rum
1/2 ounce orange curacao
1/2 ounce fresh lime juice
1/2 ounce Hawaiian Cane Syrup
1/4 ounce orgeat syrup
Cracked ice
1 ounce Lemon Hart rum (86)

In a double Old Fashioned glass combine light rum, dark rum, curacao, lime juice, Hawaiian Cane Syrup and orgeat syrup; stir well. Fill glass with cracked ice. Gently pour Lemon Hart rum on top to float on surface. Garnish with fresh pineapple wedge, vanda orchid and a sprig of fresh mint. Makes 1 drink.

PINEAPPLE COOLER

1 quart pineapple juice
2 cups apple juice
1/2 cup fresh lemon juice
1 quart pineapple sherbet
Sprigs of fresh mint

Blend together pineapple juice, apple juice and lemon juice and chill in the freezer until ice crystals start to form. Place glasses in freezer to become frosted. When ready to serve place scoops of pineapple sherbet in individual glasses and pour in the chilled juices. Garnish with sprigs of mint. Serve immediately. Makes 12 servings.

THE ROYAL PINEAPPLE

1 small fresh pineapple
1 1/2 ounces light rum
3 ounces pineapple juice
1/2 ounce fresh lemon juice
1 teaspoon Hawaiian Cane Syrup
Cracked ice

Cut off the top of the pineapple about 1 inch below the crown. Hollow out the cavity, leaving 1/2 inch of the fruit inside the rind. Reserve fresh fruit for other uses. Cut a notch in the crown. Blend together rum, pineapple juice, lemon juice and Hawaiian Cane Syrup. Fill the pineapple container with cracked ice and pour in the rum mixture. Top pineapple with the crown and insert two straws through the notch. Makes 1 drink.

RECIPES – Salads

Shrimp and Pineapple Salad with Tarragon Dressing.

SHRIMP AND PINEAPPLE SALAD

*lbs. large shrimp
(26-30 count)
lb. asparagus, large spears,
trimmed 4 to 5 inches long*

Cook shrimp, remove shells, leaving tails on, and chill. Cook asparagus 5 to 8 minutes, until tender-crisp. Drain and Chill.

*Leaf lettuce, for garnish
1/2 lbs. shredded iceberg
lettuce
hard-cooked eggs, halved
4 canned pineapple slices
lemons, cut in sixths
2 parsley sprigs
quart Tarragon Dressing*

Line serving platters with leaf lettuce. Add 2 oz. (1 cup size) shredded lettuce to each. Arrange 2 to 3 oz. shrimp (about), 2 oz. asparagus spears (4 to), 1/2 hard-cooked egg, and 2 well-drained pineapple slices on each. Garnish with lemon wedge and parsley sprig. Serve Tarragon Dressing on the side. Makes 12 servings.

HAWAIIAN REUBEN SALAD

*lbs. leaf lettuce
lbs. iceberg lettuce, shredded
lbs. corned beef strips
lbs. Swiss cheese strips
lbs. pineapple chunks in
heavy syrup, drained
lbs. 8 oz. sweet-sour red
cabbage, drained well
6 parsley sprigs
quart Thousand Island or
Louis Dressing*

For each serving line bowl with leaf lettuce (about 2 1/2 oz.). Add 4 oz. shredded lettuce (2 cups). Arrange 2 1/2 oz. corned beef and 2 oz. cheese on either side of 4 oz. pineapple chunks. Arrange 3 1/2 oz. cabbage (1/2 cup) between beef and cheese. Garnish with parsley sprig. Serve with 2 oz. Thousand Island or Louis dressing. Makes 16 servings.

LUAU FRUIT SALAD

*1 can (20 oz.) pineapple chunks
   in juice
3 oranges, peeled and sectioned
2 apples, cored and chopped
1 papaya, peeled, seeded, and
   cut into chunks
Pineapple-Poppy Seed Dressing,
   recipe follows
5 quarts torn romaine lettuce
1/2 cup slivered almonds,
   toasted*

Drain pineapple; reserve juice for dressing. Combine pineapple, oranges, apples and papaya. Toss fruit with Pineapple-Poppy Seed Dressing and refrigerate. Just before serving, toss with lettuce and sprinkle with almonds. Makes 12 servings.

PINEAPPLE-POPPY SEED DRESSING

*Reserved pineapple juice from
Luau Fruit Salad, above
1 teaspoon cornstarch
1/4 cup vegetable oil
2 tablespoons sugar
2 tablespoons distilled
   white vinegar
1 tablespoon poppy seeds
1 teaspoon grated orange peel
1/2 teaspoon paprika
1/2 teaspoon salt*

In a saucepan, combine reserved pineapple juice and cornstarch. Cook, stirring, until mixture boils and thickens. Cool. Turn into blender with oil, sugar, vinegar, poppy seeds, orange peel, paprika, and salt. Stir until blended.

RECIPES – Entrees

Sweet and Sour Pork.

SWEET AND SOUR PORK

1 pound boneless pork loin
2 tablespoons distilled white vinegar
2 tablespoons soy sauce
1 tablespoon cornstarch
2 teaspoons Asian sesame oil
1 egg yolk
1/2 cup cornstarch
1/2 cup all-purpose flour
2 cups vegetable oil
2 cloves garlic, pressed or minced
1 large onion, chopped
1 green bell pepper, seeded and cut into chunks
1 cup sliced carrots
Sauce:
1 can (20 oz.) pineapple chunks in syrup
1/4 cup catsup
1/4 cup water
2 tablespoons sugar
2 tablespoons distilled white vinegar
1 tablespoon each: cornstarch and soy sauce
1/2 teaspoon Asian sesame oil

Trim fat from pork; cut meat into one inch cubes. Combine vinegar, soy sauce, cornstarch, and sesame oil; pour over pork and marinate 30 minutes. Drain. In a small bowl, whisk egg yolk; mix in pork. In shallow dish, combine cornstarch and flour; coat pork with mixture. In large skillet, heat oil to near smoking. Brown pork in hot oil in batches (don't crowd skillet). Drain pork on paper towels. Drain off all but 2 tablespoons oil from skillet. Saute garlic, onion, green pepper, and carrots in skillet.

For sauce, drain pineapple, reserving syrup. Combine syrup with remaining ingredients. Add sauce to vegetables with pineapple. Cook 2 minutes until sauce boils and thickens. Return pork to pan; heat through. Serve with hot fluffy rice. Makes 4 servings.

SKILLET HERBED CHICKEN

1 frying chicken (2 1/2 lb.), split in half
2 tablespoons soft butter
1/8 teaspoon grated lemon peel
1/8 teaspoon salt
1/8 teaspoon tarragon, crumbled
1/16 teaspoon mint flakes, crumbled
1 (8 1/4 oz.) can pineapple chunks
1/4 cup syrup from pineapple
2 tablespoons water
1 1/2 tablespoons lemon juice
1 tablespoon prepared mustard
1 chicken bouillon cube, crumbled
1/16 teaspoon white pepper
2 teaspoons cornstarch

Loosen skin of chicken with fingers, working from breast side. Cut off wing tips. Blend butter with lemon peel, salt and herbs. Spread on the meat under skin, and skewer wing to breast. Heat a 9 inch skillet over moderate heat, and place chicken skin-side down in the dry pan. Cover and cook over medium heat 25 minutes, or until chicken is nicely browned and tender. Meanwhile, drain pineapple saving 1/4 cup syrup. Combine the 1/4 cup syrup with water, lemon juice, mustard, bouillon cube, white pepper and cornstarch for sauce. When chicken is well browned, turn pieces skin-side up. Add the drained pineapple chunks and sauce mixture to skillet. Heat to boiling, stirring gently, and spooning sauce over chicken. Simmer uncovered 5 minutes until pineapple is thoroughly heated and sauce thickens. Cut chicken pieces in half again to serve. Makes 4 servings.

ISLAND SHRIMP AND PINEAPPLE CURRY

2 cups chunked pineapple
2 lbs. fresh large shrimp
2 tablespoons butter
1 medium onion, cut in half and thinly sliced
4 cloves garlic, minced
1 tablespoon finely grated ginger
1 cinnamon stick, 2 inches in length
2 thin strips lemon zest
2 teaspoons chili powder
2 teaspoons ground coriander
1 teaspoon curry powder
2 cups coconut milk
Lemon juice, salt, rice pilaf, pickled bananas

Shell and devein shrimp, leaving tails intact. In a heavy saucepan, melt butter and saute onion and garlic until onion is translucent. Add pineapple, ginger, cinnamon, lemon zest, chili powder, coriander, curry powder and coconut milk; blend well. Bring mixture to a boil, stirring constantly, then reduce heat and simmer for 10 minutes. Add shrimp and cook for 10 to 15 minutes. Add salt and lemon juice to taste. Serve with hot rice pilaf and pickled bananas. Makes 4 to 6 servings.

RECIPES – Desserts

Pineapple Upside-Down Cake.